Pictures, Words, and Numbers

Learning Number Sense and Early Math Skills Through Page-Finding

Jillian C. Strayhorn
Joseph M. Strayhorn, Jr.

Psychological Skills Press
Wexford, Pennsylvania

Copyright © 2011 Joseph M. Strayhorn, Jr.

Psychological Skills Press

www.psyskills.com

authors' emails: jillianstrayhorn@gmail.com, joestrayhorn@gmail.com

All photos are in the public domain.

Photo credits: Thanks to pdphoto.org, public-domain-photos.com, freephoto.com, Jon Sullivan, Anna Cervova, Linda Allardice, Petr Kratochvil, Rita Ballantyne, Shari Weinsheiner, Peter Griffin, Maita Ru, Scott Meltzer, Douglas Gray, Josee-Holland Eclipse, Vera Kratochvil, Andrea Schasthuizen, Robert Kraft, Jani Ravas, Michael Meilen, Максим Кукушкин, and Bobby Mikul and other photographers providing their photos on publicdomainpictures.net.

Many thanks to Jamie Cross for his help in the design of this book.

ISBN: 978-1-931773-15-7

Introduction

This doesn't look like a math book, does it? But we constructed it with the the goal of increasing "number sense" in young people. One of the uses of this book is to teach the skill of "page-finding." Page-finding means that when someone says a page number, for example page 25, the student can open this book and fairly quickly land on the page with the picture of the waterfall and the big "25" at the bottom.

Fluent page-finding involves several mathematical skills. Let's imagine that someone hands you this book and asks you to find page 25. First, you must be able to translate between spoken sounds like "twenty-five" and numerical symbols like 25. Second, when you open the book, let's say on page 97, you have to be able to decide whether 97 is greater than 25 (in which case you move to the left) or less than 25 (in which case you'd move to the right). This is much more difficult than simply counting by rote – you have to develop a sense of where each number is relative to all the others.

But there's more: you have to develop an intuitive system of subtracting, to know whether to flip lots of pages or just a few. For example, if I'm looking for 25, and I land on 21, I only have to move a few pages forward. But if I land on page 97, I have to move lots of pages back. How do I know this? Even if I haven't studied subtraction yet, there must be some ability to know that 97 is farther from 25 than 21 is. We believe that practice in page-finding exercises this intuitive skill. And there's even more: a fluent page-finder, when asked to find page 25 in a 125 page book, will open the book somewhere close to a fifth of the way through the book, rather than, for example, near the back of the book. Intuitive mathematics is helping the student do this a very long time before the student could do this formally by a) dividing to see that 25 is a fifth of 125, b) measuring the thickness of the book, c) multiplying the thickness by one-fifth, and d) opening at the thickness indicated by these operations. In other words, fluent page-finding involves some sort of intuitive use of ratio and proportion.

Thus, if our guesses are correct, teaching children to become fluent page-finders will give them a nice head start in learning mathematics. Put another way, we hypothesize that teaching page-finding skill increases students' "number sense."

There's a second reason why we became interested in page-finding. We've been pioneering methods of tutoring students by telephone. When the tutor and the student connect by phone rather than by traveling to meet each other, it becomes possible to hold sessions much more frequently and easily. They each have the same book, and they get on the same page. If the learner isn't already fluent at page-finding, then learning this skill makes telephone tutoring much easier, while at the same time giving practice in fundamental math skills. This book is designed to make it possible to teach page-finding even when the tutor and student are connected only by phone.

The first step in the learning process is for the child to become familiar with the words for the pictures. Many children will have already accomplished this before starting to learn page-finding. This step is ideally carried out in person, when the child is quite young. You just look at pictures together and say the names of the objects. If you model pointing to pictures and gleefully saying something like, "It's a waterfall!" the child will likely follow the model and name whatever pictures he or she knows. You can flip around to the pictures in any order. An extra twist on this activity can be to not only name the picture, but also name the page number. So you point to first the number, and then the picture and say some like, "Page 25 ... It's a waterfall!" Very young children can get familiar with numerical symbols though this sort of activity.

In this activity, the adult is not asking the child what the pictures are, but is prompting the child to name them by modeling. It's no crime to say, occasionally, "What's that?" and let the child name the picture. But you want to have the activity be fun and not have the child feel that he or she is being grilled for information, especially if the child doesn't know the answers.

Once the child knows the names of all the pictures, a tutor can use the pictures to make sure the learner is on the same page in telephone tutoring.

The basic progression for teaching page-finding is as follows.

1) Selecting a set of pages to work with. This can be as small as the first 5 pages, or the entire set of 125, depending upon the student's prior skill.

2) Going through those pages in order, with the tutor or child naming the pictures, and the tutor naming the numbers. For a telephone tutor, this sounds like: "What do you see on the next page?... Yes, the moon! That's on page number 5! Do you see the 5 at the bottom of the page? Great!"

3) Going through the pages in order, with the student naming the object and the number at the bottom of the page. This may sound like this:
Tutor: How about the next page, what do you see?
Student: A waterfall!
Tutor: Yes, and what's that page number, at the bottom?
Student: 25!
Tutor: You got it! How about the next?

If the student already has picked up some counting knowledge, the student may want to flip faster and look at the numbers and name them without naming the pictures. The telephone tutor should just stop every 10 pages or so and check, with the pictures, to make sure that the child is looking at the correct numbers as he counts.

Another variation on this activity is to name the pictures and numbers, or just the numbers, in order of highest to lowest rather than lowest to highest. This will check to see if the student can really recognize the numbers without relying upon rote counting memory.

4) After going through the pages in order enough times, the student may be ready for page-finding. The tutor picks a page at random in the set and says, "See if you can find page ___." If necessary, the tutor says something like, "25 has 2 on the left and 5 on the right." When the student says, "I've found it," the phone tutor might say, "What do you see on that page?" When the student names the correct picture, the tutor says, "You got it! Yay!"

5) This series of steps is repeated for larger and larger sets of pages. For example, a student starts with pages 1 through 10, then does 11 through 20, then merges the two sets to page-find with 1 through 20. The tutor enlarges the sets slowly enough that the student gets the answers right, and fast enough that the process is not boring.

If you can use these steps to teach your student page-finding, you will have given him or her an important and valuable gift!

If you want to continue to use this book as a "math manipulative" after the student has mastered page-finding, you can also use it to model addition and subtraction. For example: "Let's start at page 5. Move 4 pages to the right. What page do you land on? 9 is right, you've added 4 and 5! 4+5 equals 9!" To subtract, you just go to the left rather than the right. "Let's start at page 9. Move back 3 pages to the left. What page do you land on? Yes, 6! You've subtracted 9-3, to get 6!"

If you want to add or subtract 3, you start counting "1" on the page just to the right or left of the page you're on, not with the page you're on. That can be tricky, and it may need to be explicitly explained and modeled. Also, when you add 3, you're saying "1, 2, 3" on pages that have other numbers on them. That's tricky too. You may have to explain that you have to ignore the page numbers from the time you start counting until you finally land.

We hypothesize that a book with interesting pictures and big page numbers, in the hands of a curious student and a joyous tutor, is a great way to teach and learn early math skills. If your observations and experience shed light on whether this hypothesis is correct, please let us know. Our email addresses are on page ii!

Cherries

1

Palm trees

Acorns

The moon

Penny

Clouds

Rose

Big bridge

Tea cup

9

White pencil

Strawberry

Hot air balloon

Leaf

Fish

Bird feeder, bird

Clock

Candle flame

Flowers, mountain

18

Camel

19

Fireworks

20

Cactus

Merry-go-round horse

Fountain

23

Waves on rocks

Waterfall

Airplane

26

U.S. flag

Toy bears

Hiker on mountain

Book

Guitar

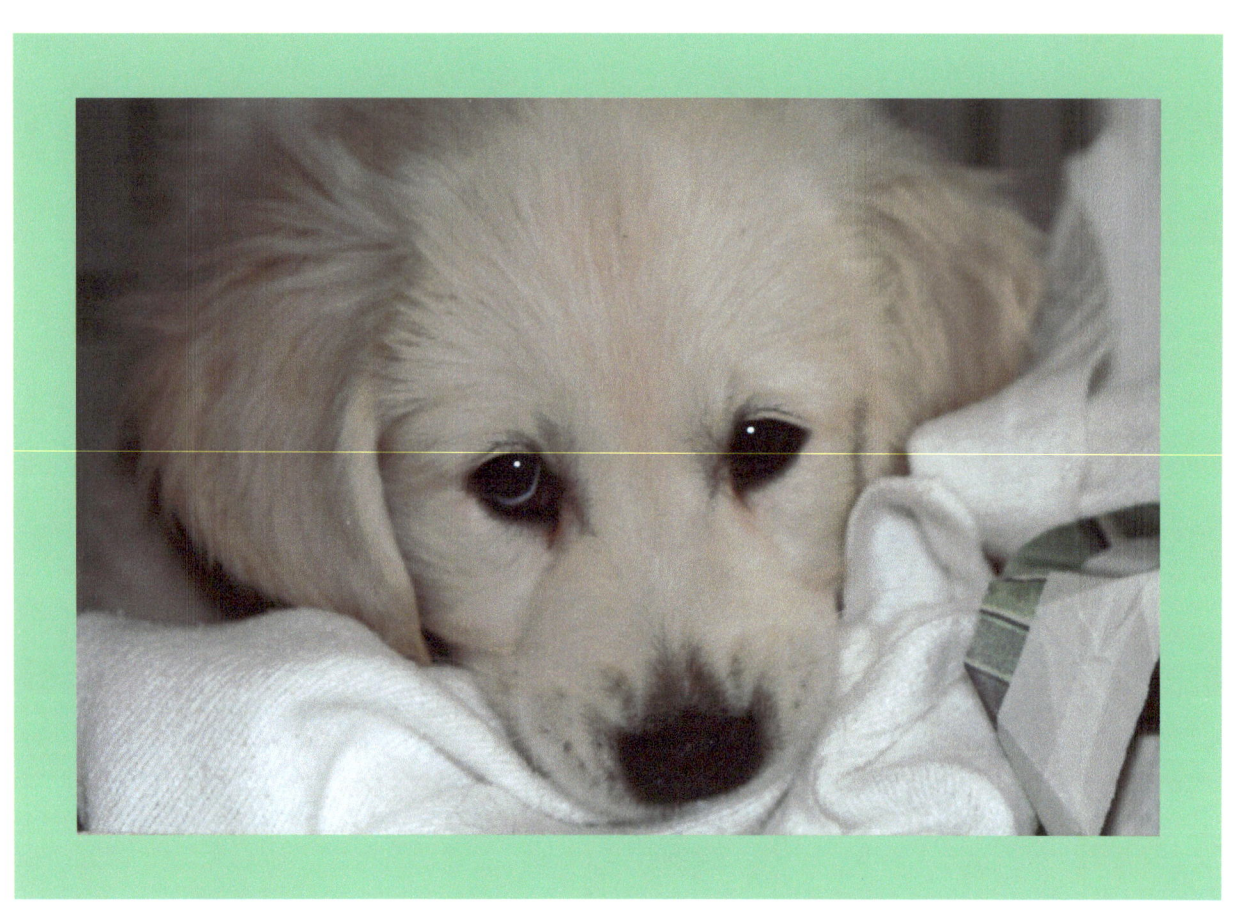

Puppy

32

Skier

Shell

Hands

35

Swan

36

Bananas

Jack-o-lanterns

Spaghetti with sauce

Paint, brush

Snail

Pizza

Cookies

Dandelion

Eggs in nest

Road

Toothbrushes

Peanuts

Rake

Notebook, marker

Glasses

Bowl of cereal

Stones

Water drop

Grass

Paper clips

Frog

Rings

Star

59

Hundred dollar bill

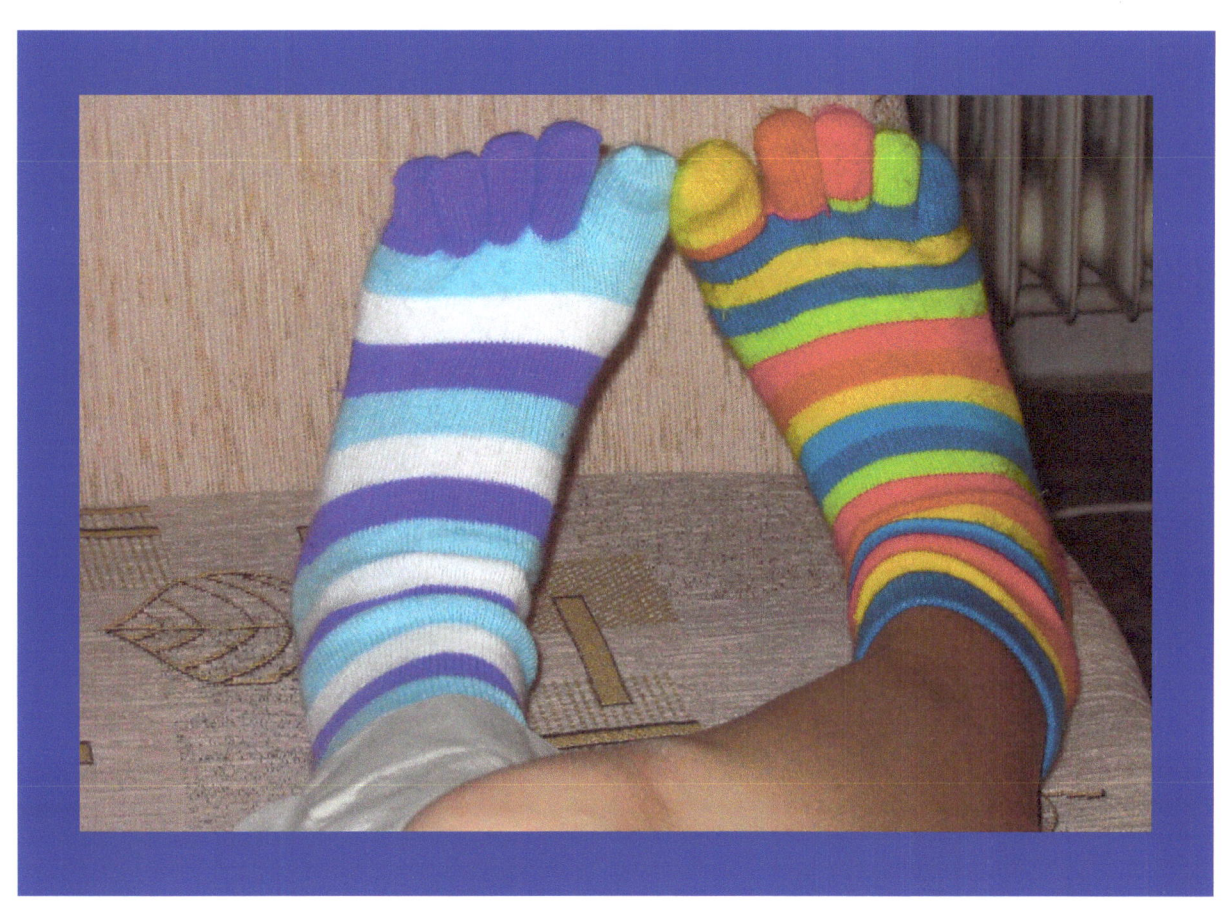

Feet with socks

Lady bug

Bike

City

64

Feather

Fan with light

Park bench

Stop sign

Bottle

Helicopter

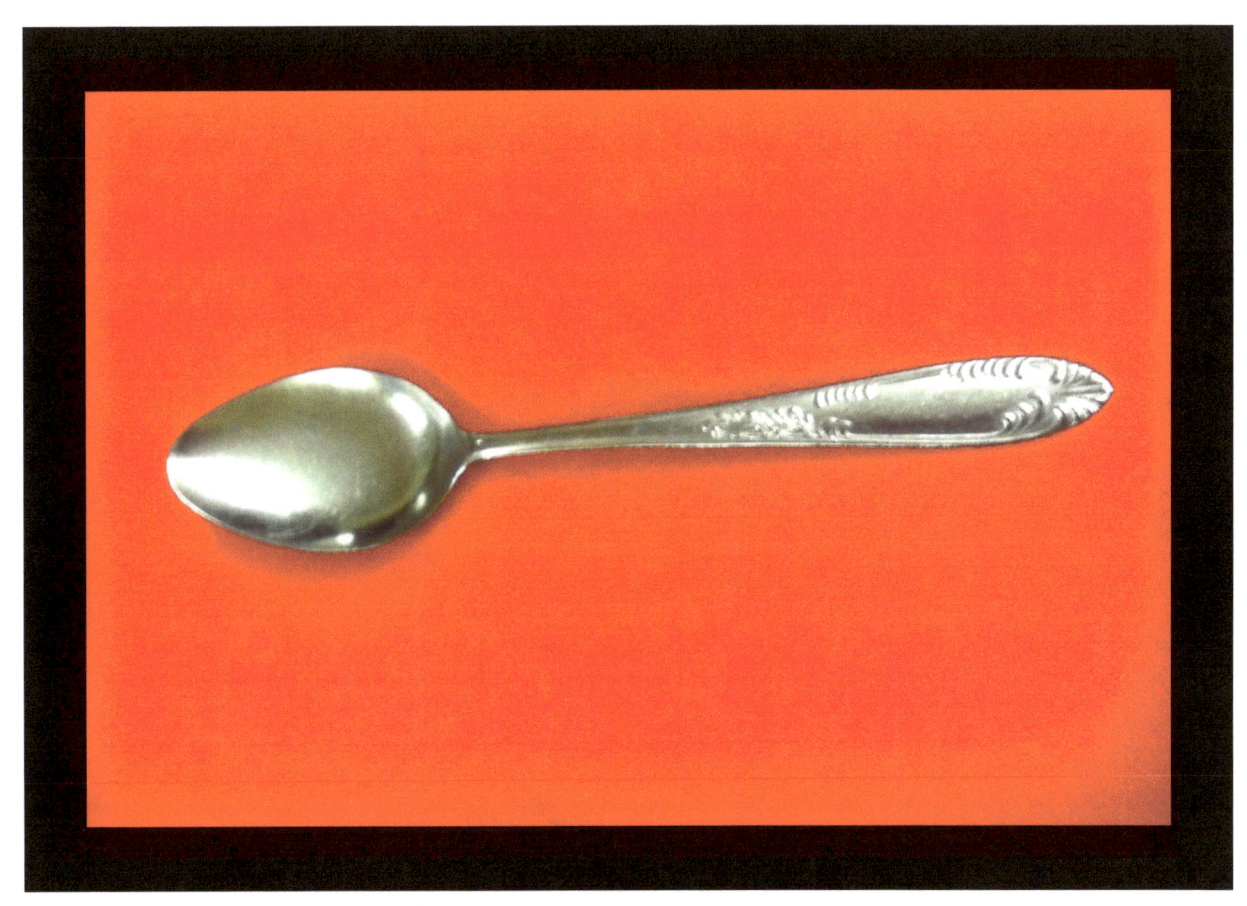

Spoon

Pair of shoes

Slide

Sink

Deck of cards

Sports car

Pair of gloves

Fan

Sailboat

Mailbox

80

Rubber ducks

Swings

82

Shopping carts

Towel

84

Tire

House

86

Escalator

Computer mouse

88

Colored lights

Soda can

Watering can

Garden

Train

Gumball machines

Balloons

Rides

Statue

Tic-tac-toe

Pot

Toy panda

Bedroom

101

Wheelbarrow

Headphones

Present

Lightning

Bell

Quilt

Railroad tracks

108

Tents

Camera

110

Tissue box

Toy car

Snowmen

113

Peas

Sheep

Ice skates

Earring

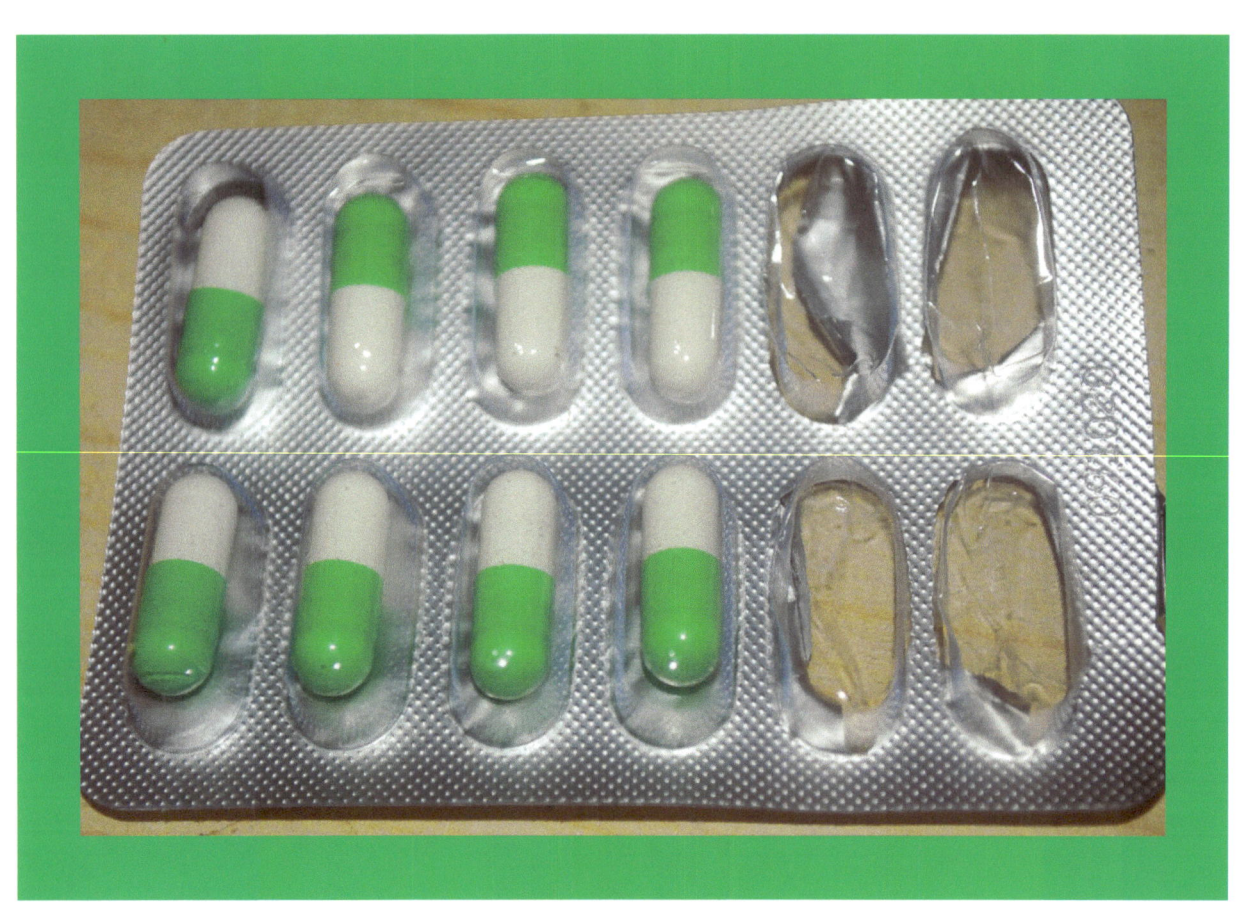

Pills

Hats

Small bridge

Lion

Hairbrush and comb

Smoothie drinks

Rainbow

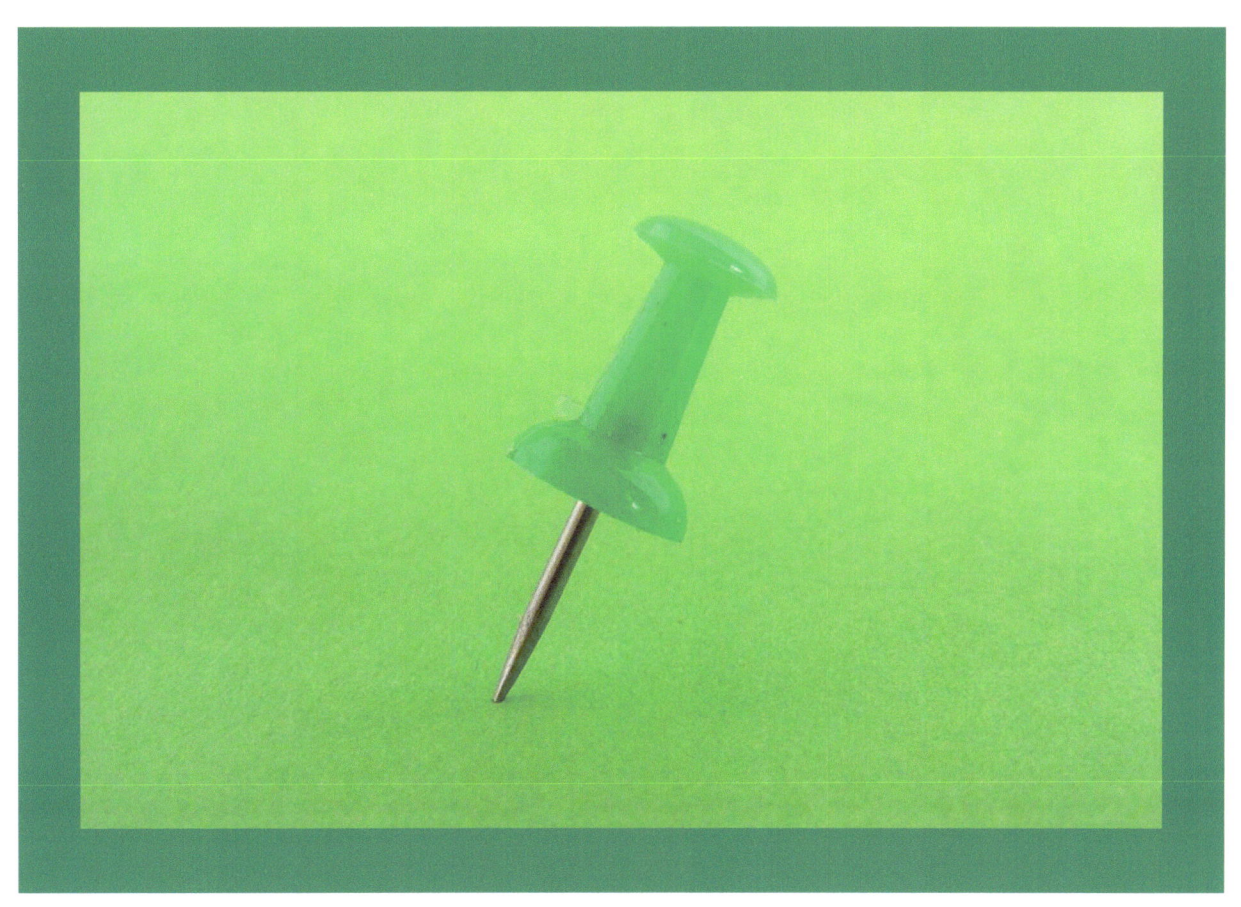

Pin

125

www.ingramcontent.com/pod-product-compliance
Lightning Source LLC
Chambersburg PA
CBHW040905020526
44114CB00037B/60